BIODIVERSITY IN THE NORTH WEST

Other titles by Bruce Ing
from the same publisher

*Biodiversity in the North West:
The Mildews of Cheshire,
Lancashire and Cumbria* (2020)

*The Fungi of North East Wales:
A Mycota for Vice-Counties 50 (Denbighshire)
and 51 (Flintshire)* (2020)

*Biodiversity in the North West:
The Slime Moulds of Cheshire* (2011)

*The Exciting World of
the Slime Moulds* (2008)

BIODIVERSITY IN THE NORTH WEST

The Slime Moulds of Lancashire and Cumbria

Bruce Ing

University of Chester Press

First published 2020
by University of Chester Press
University of Chester
Parkgate Road
Chester CH1 4BJ

Printed and bound in the UK by the
LIS Print Unit
University of Chester
Cover designed by the
LIS Graphics Team
University of Chester

© University of Chester, 2020
Cover image and some illustrations
© the named copyright holder

The moral right of the author
of this work has been asserted

All Rights Reserved
No part of this publication may be reproduced,
stored in a retrieval system or transmitted in any
form or by any means without the prior permission
of the copyright owner, other than as permitted by
current UK copyright legislation or under the terms
and conditions of a recognised copyright
licensing scheme

A catalogue record for this book is available
from the British Library

ISBN 978-1-910481-06-6

CONTENTS

Acknowledgements	vi
List of illustrations	vii
List of abbreviations	viii

PART 1

Introduction
1.1	The biology of slime moulds	1
1.2	The counties of Lancashire and Cumbria and their natural characteristics	5
1.3	The study of slime moulds in Lancashire and Cumbria	7
1.4	List of individual collectors	10
1.5	Sources of records	11
1.6	Layout of entries in the systematic account	11
1.7	Bibliography	12

PART 2

Systematic account The Slime Moulds of Lancashire and Cumbria	17
Illustrations	69
Index to genera	75

ACKNOWLEDGEMENTS

I wish to thank all those individual collectors who, over many decades, have made their collections and records available to me. Special thanks are due to John Watt, Stuart Skeates and Peter Smith for help with databases and records. The curators on national herbaria have been of considerable help in providing me with access to historic material and libraries across the region have helped me to locate obscure local literature. Thanks also to Sarah Griffiths and her colleagues at the University of Chester who have been very patient with me! I am most grateful for the companionship of colleagues on forays organised by the British Mycological Society and the North West Fungus Group, but above all, to my wife Ellie, who has given patient support to my endeavours for over fifty years and is still smiling!

LIST OF ILLUSTRATIONS

Figure 1: The vice-counties of Lancashire and Cumbria — 69

Figure 2: *Arcyodes incarnata* – an uncommon species on sticks in damp places (John Watt) — 70

Figure 3: *Arcyria obvelata* – a common species on fallen trunks, especially beech (John Robinson) — 71

Figure 4: *Lepidoderma tigrinum* – an uncommon species of wet, mossy rocks and moss and lichen on rotting conifer trunks (Michel Poulain) — 72

Figure 5: *Licea biforis* – a relative newcomer to our islands, arriving from the tropics as a harbinger of climate change (John Robinson) — 73

Figure 6: *Macbrideola cornea* – a common species, especially in the west, on mosses on the bark of living trees (Diana Wrigley-Basanta) — 74

LIST OF ABBREVIATIONS

BM – Herbarium of the Natural History Museum, South Kensington, London

BMSF – recorded on a foray of the British Mycological Society

E – Herbarium of the Royal Botanic Garden, Edinburgh

K – Herbarium of the Royal Botanic Gardens, Kew

MANCH – Herbarium of the Manchester Museum

MANX – Herbarium of the Manx Museum, Douglas, Isle of Man

NWFG – recorded on a foray of the North West Fungus Group

v.c. – vice-county or vice-counties

PART 1

INTRODUCTION

1.1 The biology of slime moulds

The organisms collectively known as slime moulds share some fungal and some animal, or rather protozoan, characteristics. They have amoeboid and/or flagellate single-celled feeding stages and usually some form of aggregation into either a multicellular structure – the *pseudoplasmodium*, or a multinucleate single cell – the *plasmodium*, which migrates from the feeding area of the substrate to a region where it can produce and disperse spores.

There are five major groups of slime moulds – protostelids, dictyostelids, ceratiomyxids, myxomycetes and acrasids. The first four are branches of the same evolutionary line while the acrasids have evolved from a very distant ancestor. The first four are now recognised as members of the Amoebozoa but have been classified as fungi in the past. However they are not slimy and they are not moulds! They have been called 'fungus-animals' or Mycetozoa, a term which is coming back into fashion.

The protostelids are the smallest, and simplest; some have flagellate cells and others do not. They mostly have a small multinucleate plasmodium that may be the product of a sexual process. The fruiting body is usually a thin, hollow stalk surmounted by a cluster of a few spores. Protostelids live on the outside of dead plant stems, flower heads and the bark of living trees. They can only be seen when these substrates are cultured in moist chambers. The group has been little studied in Britain.

Dictyostelids, or cellular slime moulds, are probably derived from non-flagellated protostelids but differ in several important ways. Neither sexual reproduction nor plasmodia occur. Instead, individual amoebae move towards the source of a chemical attractant, *acrasin*, and form the aggregation phase or pseudoplasmodium. This transforms into a slug-like *grex* and moves towards a drier and higher part of the substrate. Spore production involves a division of labour in which a cellular stalk is formed from some of the amoebae while those which secrete the most acrasin crawl up the stalk and form resting spores. Dictyostelids are primarily soil organisms but are also common on herbivore dung. As with all the other groups of slime moulds they feed by ingesting bacteria, other protozoans, algae or yeast fungi. They are perhaps more common in the tropics than in temperate regions and have received only moderate attention in Britain.

Ceratiomyxids are closely related to the protostelids but differ in having massive plasmodia, often a few centimetres across, and a well-developed sexual phase, with fusion of compatible cells to form the plasmodium. The fruiting body, which is conspicuous to the naked eye, consists of finger-like processes carrying very small sporangia, which in the past were interpreted as spores. They have traditionally been

Introduction

included in the myxomycetes. There are five species in the single genus, *Ceratiomyxa*, of which three occur in Britain. Only the most common, *C. fruticulosa*, occurs in our area. This is found on rotten wood, especially of conifers, and is truly cosmopolitan.

The myxomycetes, or plasmodial slime moulds, are much more complex. The plasmodial stage is often large and brightly coloured, and is capable of visible movement across the substrate. The fruit bodies often resemble miniature mushrooms, usually only a millimetre or so high, and may also be brightly coloured. As a group they occur in all climates and in all kinds of vegetation, from deserts to alpine snowfields, but are most common in forests, occurring on fallen wood of all kinds, leaf litter, plant remains, mosses, stems of living plants and the bark of living trees. The last habitat contains a wealth of minute species, barely visible to the naked eye, and usually only found by culturing pieces of bark from living trees in moist chambers – usually Petri dishes. Many species are confined to the tropics and a larger group are only known from vegetation close to melting snow on high mountains in the spring. There are almost a thousand described species of which 391 have been found in Britain.

The acrasids are not related to other slime moulds and have a quite different amoeboid stage. A flagellate stage has been reported from one species. Sexual reproduction has not been observed. The plasmodial stage is inconspicuous. Only one species is common,

Pocheina rosea, which grows on acid bark. This is bright pink and has a stalk made up of cubical cells and could be mistaken for a member of the myxomycete genus *Echinostelium*, which also lives on bark but does not have a cellular stalk, just a hollow tube. Other acrasids are found on dung, plant litter and old wine casks, but are generally regarded as rare. It is more likely that they are just insufficiently studied, especially in Britain.

    Myxomycetes are abundant on bark, even in cities, and in soil of all kinds, including deserts, where they may be important agents in the phosphate cycle. They are especially common on dead stems of nettles and rosebay willowherb, which are typically found on phosphate-rich soil. Forest litter is also rich in species and the fruit bodies may be seen on the surface. Dead wood is an important habitat and some of the larger species may feed inside a fallen trunk for several years before emerging, often through a beetle bore hole, to produce spores at the surface. Spores are mostly wind dispersed but some species appear to have formed a partnership with certain beetles, whose complete life cycle occurs within the large clusters of fruiting structures on the wood. The female insects carry spores in pouches in their jaws and have been observed 'sowing' spores on rotten wood. For more information on the biology and identification of myxomycetes see Ing (1983, 1994, 1999, 2008, 2011, 2020) and Stephenson & Rojas (2017).

Introduction

1.2 The counties of Lancashire and Cumbria and their natural characteristics

Which area is included?
The area under consideration comprises the historical counties of Lancashire, Westmorland and Cumberland. The designation of the current counties of Lancashire and Cumbria has involved radical boundary changes in recent years. Lancashire has lost the boroughs of Widnes and Warrington to Cheshire, the rest of the south to Merseyside and Greater Manchester and the Furness region to Cumbria. Westmorland has been completely taken into Cumbria which has also subsumed a small area that was in Yorkshire, in the Sedbergh and Dent area.

Naturalists prefer continuity of recording areas and use a system invented in the 1850s to allow the same areas to be compared over time. The unit of area is the vice-county (v.c.), which mainly corresponds to the existing county in 1852, with the larger counties, such as Yorkshire, being split into up to five more uniform units. Using this system we have v.c. 59 to designate the County Palatine of Lancashire, south of the Ribble, v.c. 60 for the region north of the Ribble to the old Westmorland border, v.c. 65 for Sedbergh and Dentdale, v.c. 69 for Westmorland and Furness and v.c. 70 for 'old' Cumberland (see Figure 1 for a map of these vice-counties).

The Slime Moulds of Lancashire and Cumbria

The physical and ecological environment
The geology of the area is relatively simple in the south but becomes more complex as one travels north to the Lake District. The Triassic sandstone rocks of Cheshire continue in to south Lancashire as far as the Ribble but are replaced in the east by the Carboniferous strata of the Coal Measures of the Lancashire coalfield, and then the Millstone Grit of the Pennines. North of the Ribble the sandstones make up the Fylde area but north of Preston the Millstone Grit and then Carboniferous limestone begin to dominate the scene. North of Lancaster and along the Furness coast the limestone produces moderate hills, scars and cliffs, with some fine examples of pavement around Morecambe Bay. North of this the Silurian rocks replace the Carboniferous before the complex Lake District geology takes over. This region is surrounded by a ring of limestone, enclosing Ordovician lava and slates. The younger rocks, including Coal Measures, recur only along the coast and the Triassic sandstones re-appear up to the Scottish border.

Topographically the western edges of the region are low-lying and relatively flat, but to the east are the rolling hills of the Pennine chain which continues along the Pennine edge towards the Scottish border, confronting to the west the mountains and lakes of the Lake District.

The flat lands north of the Mersey are punctuated with the remains of peat mosses and these recur along the coast up to the Scottish border, and beyond. The

Introduction

Pennine hills are important for their sessile oakwoods and the limestone provides grassland, woodland and pavements, all of which support a wide range of microhabitats. The Lake District supports fine Atlantic woods in Borrowdale and around Buttermere, which are very rich in myxomycetes. Fine stretches of sand dunes, with grassland and woodland, such as Ainsdale and Ravensdale are important conservation areas, as are the Cliburn, Meathop and Roudsea mosses and Eaves Wood, Silverdale and Roudsea Wood. Much of the hinterland is arable and pasture, with old estates with amenity woodland and gardens, all of which yield substrates for our organisms.

1.3 The study of slime moulds in Lancashire and Cumbria

The earliest reference is from the Rev. H.H. Higgins (1859) who collected in the Liverpool area. H. Murray collected in the Manchester area between 1899 and 1906 and his material is in the Manchester Museum (**MANCH**). Two well-known Merseyside mycologists, Dr J.W. Ellis and H.J. Wheldon, made many collections in the years leading up to World War I and their material is at **K** and **BM**. Another renowned worker, Dr E.M. Blackwell studied the fungi of the coastal dunes north of Liverpool between 1916 and 1919, and recorded several myxomycetes. W.G. Travis also recorded in the south of the county in the 1920s. After World War II, J. Terry Palmer collected widely, especially in the dunes and W.S. Lacey recorded several myxomycetes around Chorley. J.E. Milne, who lived at Bramhall, also collected widely in the south of the county, until he moved to the Isle of Man. His collections are in the Manx Museum in Douglas (**MANX**). Local natural history societies such as the Liverpool Botanical Society, Liverpool Field Naturalists, Merseyside Naturalists and Chorley and District Natural History Society all began to have fungus forays at this time and a number of amateur mycologists started to record myxomycetes. Prominent among them was Pat Livermore, in Lancaster, who made numerous records in Gait Barrows and surrounding areas. More recently Peter Smith of Bolton, a very active member of the North West Fungus Group, began to specialise in myxomycetes and the results of his labours are well seen in the systematic account in Part 2. The present author came to Lancashire for the 1965 foray of the British Mycological Society and has been collecting and recording in the North West ever since.

There has been much less activity in Westmorland and Cumberland. The earliest published records were in the *Westmorland Notebook* of 1889, by C.H. Waddell and J. Atkinson. Many records were accrued during the several forays organised by the British Mycological Society (see below). The present author, while working for the Council for Nature, started visits to many sites in the Lake District and Furness in 1961 and is still a regular visitor to the area. George

Introduction

Massee, one of the Victorian 'greats', listed numerous species from Carlisle in his 1892 *Monograph of the Myxogastres*, but without specific localities; his material, equally sparsely labelled, is in **K** and **BM**.

The British Mycological Society has organised residential forays in the region at regular intervals, as follows: 1922 (Keswick), 1954 (Far Sawrey), 1962 (Grange-over-Sands), 1965 (Manchester), 1971 (Windermere), 1972 (Liverpool), 1984 (Morecambe), 1985 (Chester), 1986 (Murton, Appleby), 1992 (Ambleside), 1995 (Murton), 1999 (Blencathra), 2002 (Ashton Hall, near Preston), 2006 (Carnforth). The present author was present from 1965 to 2002 at these forays. Since that time the North West Fungus Group has run regular day forays throughout the year and held residential weekends at Keswick.

The total number of species recorded from each of the vice-counties (or part v.c. in the area, is as follows: 59 (South Lancashire) **160**; 60 (North Lancashire) **125**; 65 (North-west Yorkshire in Cumbria) **34**; 69 (Westmorland and Furness) **133** and 70 (Cumberland) **140**. This compares well with the more intensive survey of Cheshire which yielded 182 species (Ing, 2011).

The Slime Moulds of Lancashire and Cumbria

1.4 List of individual collectors

A. Adams; J. Adams; J. Atkinson; N. Bamforth; C.E. Barley; S.S. Bates; M.J. Berkeley; E.M. Blackwell; H. Britten; D. Carlyle; A. Carter; C.D. Cayley; C.H. Chadwick; W.N. Cheesman; M.C. Clark; G. Clarkson; P. Cook; W.F. Davidson; R. Dean; T. Dean; W.T. Elliott; J.W. Ellis; G.W. Garlick; D.V. Gelvin; M. Hall; B. Hartham; H.H. Higgins; D. Higginson-Tranter; D. Hinchcliffe; P. Holland; B. Ing; J. Jordan; S. Lanning; W.S. Lacey; T. Lassoe; G. Lister; J.J. Lister; P. Livermore; J. Maddy; A. McKernan; George Massee; J.E. Milne; D.W. Mitchell; A. Moor; H. Murray; Y. Mynett; D. Nelson; W.E. Nicholson; North West Fungus Group; T. Osborne; J.T. Palmer; R. Poole; A. Richardson; I. Ridge; R.W. Robson; J. Rose; G.S. Rowley; C.H. Sidebotham; P.R. Smith; W. Sperink; P.R. Stewart; G. Taylor; J. Taylor; F.J. Thorpe; W.G. Travis; M. Valentine; C.H. Waddell; J.T. Wadsworth; H.M. Ward; J.W. Watt; H.J. Wheldon; P.H. Woodhead; Yorkshire Naturalists' Union.

Where a collector's name does not appear in the species accounts his/her collections were of more common species. The initials *BMSF* and *NWFG* refer to collections made during British Mycological Society forays and those of the North West Fungus Group, where the name of the collector was not recorded.

Introduction

1.5 Sources of records

Records have been collated from a variety of sources. The list of references in the bibliography gives details of relevant published accounts. The herbaria at **BM, E, K, MANCH** and **MANX** have been studied. The author's extensive herbarium has voucher specimens for most of the species recorded in Lancashire and Cumbria.

The fungal database of the British Mycological Society has details of all its forays and the database of the North West Fungus Group has provided much information. Finally the account would not be complete without the records provided by the individual collectors named in the previous section.

1.6 Layout of entries in the systematic account

Currently accepted name author citation
Synonyms in recent literature
Habitat in Lancashire and Cumbria; general distribution in Britain.

Where there are fewer than six records: the vice-county, locality, date and collector, followed by the 10 km grid square.

Where there are six or more records: the relevant vice-counties are listed followed by years of first and latest records. All the recorded 10 km grid squares from which the species is known are given, then the number of sites across the region. Where appropriate, notes are added.

The nomenclature follows Ing (1999).

1.7 Bibliography

Atkinson, J. (1889) List of fungi. *Westmorland Notebook.* **1**(7), 90–94.

Cheesman, W.N. & Elliott, W.T. (1925) Report on the Mycetozoa found during the foray at Keswick. *Transactions of the British Mycological Society* **9**, 12–14.

Higgins, H.H. (1859) The fungi of Liverpool: Gasteromycetes. *Proceedings of the Literary and Philosophical Society of Liverpool* **59**, 13–38.

Holden, M. (1966) Autumn foray, Manchester, 1965. *News Bulletin of the British Mycological Society* **26**, 1–4.

Ing, B. (1983) A ravine association of myxomycetes. *Journal of Biogeography* **10**, 299–306.

Ing, B. (1994) The phytosociology of Myxomycetes. *New Phytologist.* **126**, 175–201.

Ing, B. (1999) *The Myxomycetes of Britain and Ireland.* Slough, Richmond Publ. (reprinted 2020).

Ing, B. (2008) *The Exciting World of the Slime Moulds.* Chester, University of Chester Press.

Ing, B. (2011) *Biodiversity in the North West: The Slime Moulds of Cheshire.* Chester, University of Chester Press.

Ing, B. (2020) Three new species of myxomycetes. *Field Mycology* (in press).

Lacey, W.S. (1954) Notes on the flora of the Chorley District of South Lancashire. *North Western Naturalist* (NS) **2**, 526–558.

Introduction

Livermore, P.D. & L.A. (1987) *Fungi of Gait Barrows NNR.* Bowness-on-Windermere, Nature Conservancy Council.

Massee, G. (1892) *Monograph of the Myxogastres.* London, Methuen.

Milne, J.E. (1971) Fungi, in Kidd, L.N. & Fitton, M.G. (eds.) *Holden Clough.* Oldham, Oldham Public Libraries, Art Gallery & Museum.

Minter, D.W. & Moodie, W.Y. (1985) Spring foray, 1984, Morecambe, Lancashire, 23–29 May. *Bulletin of the British Mycological Society* **19**, 17–23.

Montgomery, N. (1963) Spring foray, 24–28 May, 1962, Grange-over-Sands. *News Bulletin of the British Mycological Society* **19**, 4–6.

Palmer, J.T. (1952) Freshfield: A mycological survey for 1951. *Proceedings of the Liverpool Naturalists' Field Club* **1951**, 18–25.

Stephenson, S.L. & Rojas, C. (eds.) (2017) *Myxomycetes.* London, Academic Press.

Thomas, A. (1972) Spring foray, Windermere, 13–17 May 1971. *Bulletin of the British Mycological Society* **6**, 4–6.

Thomas, A. (1973) Autumn foray, Liverpool, 1972. *Bulletin of the British Mycological Society* **7**, 52–58.

Travis, W.G. (1925) Some South Lancashire myxomycetes. *Lancashire and Cheshire Naturalist* **17**, 87.

Waddell, C.H. (1889) List of fungi. *Westmorland Notebook* **1**(7), 61–63.

Wheldon, H.J. (1914) The fungi of the sand-dune formations of the Lancashire coast. *Lancashire and Cheshire Naturalist* **7**, 218–219.

Yates, R. (1990) Checklist of fungi in the Chorley District. *Annual Report of the Chorley and District Natural History Society* **1990**, 59.

PART 2

SYSTEMATIC ACCOUNT

THE SLIME MOULDS OF LANCASHIRE AND CUMBRIA

Phylum HETEROLOBOSA
Class ACRASIOMYCETES
Order ACRASIDALES
Family Guttulinaceae

Pocheina rosea (Cienk.) A.R. Loeblich & Tappan
On acid bark of living trees, in moist chamber culture, especially in urban areas affected by atmospheric pollution; common.
Recorded in v.c. 59, 60, 69, 70. First record 1981, latest record 2011.
SJ 48, 49, 69; SD 20, 29, 31, 38, 41, 47, 48, 50, 66; NY 22, 52, 72. 16 sites.

The Slime Moulds of Lancashire and Cumbria

Phylum MYXOMYCOTA
Class PROTOSTELIOMYCETES
Order PROTOSTELIALES
Family Protosteliaceae

Protostelium mycophaga Olive & Stoioan.
On bark of living trees; uncommon.
69. Witherslack Woods, 1981, *B. Ing*. SD 48.
70. Keskadale Oaks, 1982, *B. Ing*, NY 21.

Class DICTYOSTELIOMYCETES
Order DICTYOSTELIALES
Family Dictyosteliaceae

Dictyostelium brefeldianum Hagiwara
Dictyostelium mucoroides auctt.
On horse dung; frequent.
59. Walton, Liverpool, 1924, *W.G. Travis*, SJ 39.

Systematic Account

Class CERATIOMYXOMYCETES
Order CERATIOMYXALES
Family Ceratiomyxaceae

Ceratiomyxa fruticulosa (Müll.) T. Macbr.
On rotten wood, especially of conifers, summer to autumn; common.
Recorded in v.c. 59, 60, 69, 70. First record 1859, latest record 2019.
SJ 48, 49; SD 20, 21, 29, 31, 32, 38, 39, 41, 45, 47, 48, 50–52, 56, 57, 61, 62, 66, 73, 81, 90, 91; NY 11, 20–22, 30, 32, 35, 41, 42, 51, 72. 57 sites.

Class MYXOMYCETES
Order ECHINOSTELIALES
Family Clastodermataceae

Clastoderma debaryanum A. Blytt
On bark of living trees; uncommon.
69. Low Wood, Hartshop, 1981, *B. Ing*, NY 41.

Clastoderma pachypus Nann.-Bremek.
On bark of living oaks; frequent.
70. The Ings, Borrowdale, 1992, *B. Ing*, NY 22.

Family Echinosteliaceae

Echinostelium brooksii K.D. Whitney
On acid bark of living trees, especially conifers; common.
Recorded from v.c. 59, 60, 69, 70. First record 1986, latest record 2011.
SJ 48, 69; SD 20, 21, 38, 41, 47, 48, 50, 63, 71; NY 22, 52, 81. 14 sites.

Echinostelium colliculosum K.D. Whitney & H.W. Keller
On bark of living trees; common.
Recorded from v.c. 59, 60, 69, 70. First record 1975, latest record 2013.
SJ 49, 59, 69; SD 38, 47, 48, 51, 61, 71, 72; NY 21, 22, 41, 42, 51. 19 sites.

Echinostelium corynophorum K.D. Whitney
On bark of living trees; frequent.
Recorded from v.c. 59, 60, 65, 69, 70. First record 2008, latest record 2013.
SJ 59, 69; SD 63, 79, 72; NY 45, 60. 7 sites.

Echinostelium fragile Nann.-Bremek.
On bark of living trees; common.
Recorded from v.c. 59, 60, 65, 69, 70. First record 1975, latest record 2010.
SJ 69; SD 20, 33, 38, 47, 50, 51, 56, 61, 63, 69; NY 22, 41, 44, 52, 62. 16 sites.

Systematic Account

Echinostelium minutum de Bary
On bark of living trees; common.
Recorded from v.c. 59, 60, 69, 70. First record 1974, latest record 2019.
SJ 48, 49, 59, 69; SD 20, 31, 38, 39, 41, 47–49, 56, 61, 66, 71, 72; NY 21, 22, 41, 42, 45, 51, 62, 81. 27 sites.

Order CRIBRARIALES
Family Cribrariaceae

Cribraria argillacea (Pers.) Pers.
On rotten conifer wood; common.
Recorded from v.c. 59, 60, 69, 70. First record 1885, latest record 2002.
SJ 48; SD 21, 38, 39, 57, 73; NY 22, 35, 42, 52, 61, 72. 14 sites.

Cribraria aurantiaca Schrad.
On rotten conifer wood; common.
Recorded from v.c. 59, 60, 69, 70. First record 1884, latest record 2005.
SJ 48; SD 20, 21, 39, 47, 51, 57; NY 21, 22, 45, 52, 70, 72. 15 sites.

Cribraria cancellata (Batsch) Nann.-Bremek.
On rotten conifer wood; common.
Recorded from v.c. 59, 60, 69, 70. First record 1889, latest record 2010.
SJ 48, 89; SD 20, 21, 39, 48, 49, 57; NY 32, 35, 52, 72. 13 sites.

Cribraria intricata Schrad.
On rotten wood, especially of oak; rare.
59. Ainsdale, 1972, *B. Ing*, SD 21.
70. Carlisle (Massee, 1892). NY 35.

Cribraria minutissima Schwein.
On fallen branches of oak; rare.
59. Duxbury Wood, 2004, *P. Smith*, SD 51.
This species is otherwise known in the UK from single sites in Yorkshire and Stirling.

Cribraria persoonii Nann.-Bremek.
On rotten conifer wood; frequent.
59. Freshfield, 1960, *S.S. Bates*, SD 20; Ainsdale, 2002, *B. Ing*, SD 21.
69. Murton, 1981, *B. Ing*. NY 72.

Cribraria rufa (Roth) Rostaf.
On rotten conifer wood; common.
Recorded from v.c. 59, 60, 65, 69, 70. First record 1892, latest record 1999.
SD 57, 61, 68; NY 35, 42, 72. 6 sites.

Cribraria tenella Schrad.
On rotten wood, not restricted to conifers; uncommon.
69. Brantwood, 1961, *B. Ing*, SD 39.

Systematic Account

Cribraria violacea Rex
On bark of living broad-leaved trees; frequent.
59. Knowsley, 1988, *T. Lassoe*, SJ 49.
60. Gait Barrows, 1994, *B. Ing*, SD 47.
69. Appleby, 1994, *B. Ing*, NY 62.
70. Lodore Falls, Borrowdale, 1992, *B. Ing*, NY 21.

Lindbladia tubulina Fr.
On fallen conifers, stumps and sawdust heaps; uncommon.
60. Warton Crag, 2005, *NWFG*, SD 57.

Family Dictydiaethaliaceae

Dictydiaethalium plumbeum (Schum.) Rostaf.
On fallen trunks, especially of beech and elm; frequent.
Recorded from v.c. 59, 60, 69, 70. First record 1898, latest record 1999.
SJ 48, 79; SD 47, 50, 56; NH 42, 52. 7 sites.

Family Liceaceae

Licea belmontiana Nann.-Bremek.
On bark of living trees; uncommon.
Recorded from v.c. 59, 60, 65, 69, 70. First record 1981, latest record 2014.
SJ 79; SD 63, 73, 79, 80; NY 21, 31, 81. 8 sites.

Licea biforis Morgan
On bark of living trees; becoming more common as the climate warms.
Recorded from v.c. 59, 60, 65, 69, 70. First record 1986, latest record 2014.
SJ 59, 89; SD 47, 48, 67, 69, 73; NY 45. 8 sites.

Licea bryophila Nann.-Bremek.
On the liverworts *Metzgeria* and *Radula* on living trees; frequent.
Recorded from v.c. 59, 60, 69, 70. First record 1981, latest record 2011.
SD 39, 46, 47, 62; NY 21, 41. 6 sites.

Licea castanea G. Lister
On bark of living trees; uncommon.
59. Smithills Hall, 2006, *B. Ing*, SD 71; Haigh Hall, 2006, *B. Ing*, SD 50; Darwen, 2009, *B. Ing*, SD 62.
69. Tarn Hows, 2003, *B. Ing*, SD 39.

Licea clarkii Ing
On dead, arching stems of brambles; frequent.
Recorded from v.c. 59, 60, 69, 70. First record 1999, latest record 2015.
SD 41, 47, 49; NY 22, 43. 6 sites.

Systematic Account

Licea denudescens H.W. Keller & T.E. Brooks
On lichens on the bark of living trees; frequent.
Recorded from v.c. 59, 60, 69, 70. First record 1981, latest record 2014.
SJ 49, 69, 89; SD 29, 47, 48, 51, 73, 80; NY 21, 42, 44, 51. 14 sites.

Licea eleanorae Ing
On bark of living trees; uncommon.
59. Astley Hall. 2020, *B. Ing*, SD 51.
69. Bowness, 2003, *B. Ing*, SD 49.
This species is named after the author's wife.

Licea erddigensis Ing
On bark of living trees; uncommon.
Recorded from v.c. 59, 60, 69, 70. First record 2006, latest record 2011.
SJ 48, 79; SD 46, 61, 71; NH 45. 6 sites.

Licea inconspicua T.E. Brooks & H.W. Keller
On bark of living trees; uncommon.
59. Charnock Richard, 1997, *B. Ing*, SD 51.
60. Lord's Lot, 1984, *B. Ing*, SD 57; Roeburndale, 1984, *B. Ing*, SD 66.
69. Witherslack Woods, 1981, *B. Ing*, SD 48.
70. Keskadale Oaks, 1981, *B. Ing*, NY 21.

Licea kleistobolus G.W. Martin
On bark of living trees and climbers; common.
Recorded from v.c. 59, 60, 65, 69, 70. First record 1981, latest record 2013.
SJ 48, 49, 69, 89; SD 20, 21, 23, 39, 41, 46–51, 56, 61–63, 71, 72, 79; NY 22, 44, 45, 52, 62. 31 sites.

Licea longa Flatau
On bark of living trees; rare.
59. Salford, 2010, *B. Ing*, SJ 79.

Licea lucens Nann.-Bremek.
On *Radula* on bark of living sycamore; rare.
59. Darwen, 2010, *B. Ing*, SD 62.
Known elsewhere from a single site in Scotland and three in France.

Licea marginata Nann.-Bremek.
On bark of living trees; common.
Recorded from v.c. 59, 60, 69, 70. First record 1981, latest record 2014.
SJ 49, 69, 89; SD 38, 39, 50, 57, 72, 80; NY 21, 22, 42, 45, 51, 61, 81. 18 sites.

Licea microscopica D.W. Mitchell
On cyanobacteria on bark of living elder; common.
Recorded from v.c. 59, 60, 65, 69, 70. First record 2003, latest record 2011.
SJ 49, 59, 69; SD 41, 46–49, 51, 61, 62, 69, 71; NY 45. 13 sites.

Systematic Account

Licea minima Fr.
On bark of living trees and on fallen trunks; common.
Recorded from v.c. 59, 60, 69, 70. First record 1972, latest record 2014.
SJ 48, 89; SD 21, 56, 63; NY 22, 41, 44, 62. 9 sites.

Licea operculata (Wingate) G.W. Martin
On bark of living trees; common.
Recorded from v.c. 59, 60, 65, 69, 70. First record 1986, latest record 2019.
SJ 48, 49, 59; SD 49, 51, 53, 61, 62, 67, 69, 73; NY 45, 52. 13 sites.

Licea parasitica (Zukal) G.W. Martin
On bark of living trees; common.
Recorded from v.c. 59, 60, 65, 69, 70. First record 1975, latest record 2014.
SJ 39, 48, 49, 59, 69, 79, 89; SD 20, 27, 29, 38, 39, 41, 46–51, 53, 57, 61–63, 66, 69, 71–73, 79, 80; NY 11, 21, 22, 31, 41, 42, 44, 45, 51, 52, 72, 81. 61 sites.
This is the most common myxomycete on living trees.

Licea pedicellata (H.C. Gilbert) H.C. Gilbert
On bark of living trees; uncommon.
Recorded from v.c. 59, 60, 69, 70. First record 1992, latest record 2013.
SJ 59, 79; SD 46, 48, 62; NH 21. 6 sites.

Licea perexigua T.E. Brooks & H.W. Keller
On bark of living trees; uncommon.
59. Warrington, 2008, *B. Ing*, SJ 59.

Licea pusilla Schrad.
On bark of living trees and on fallen conifer trunks; common.
Recorded from v.c. 59, 60, 69, 70. First record 1981, latest record 2011.
SD 21, 41, 47, 52; NY 22, 45, 52. 6 sites.

Licea pygmaea (Meylan) Ing
On bark of living trees; common.
Recorded from v.c. 59, 60, 69. First record 1986, latest record 2014.
SJ 79, 89; SD 50, 51, 61–63, 71, 73; NY 71. 10 sites.

Licea scintillans McHugh & D.W. Mitchell
On bark of living trees; rare.
59. Accrington, 2013, *B. Ing*, SD 72.

Licea scyphoides T.E. Brooks & H.W. Keller
On bark of living trees; frequent.
Recorded from v.c. 59, 60, 65, 69, 70. First record 1981, latest record 2010.
SJ 79; SD 49, 63, 69; NY 21, 51, 52. 7 sites.

Systematic Account

Licea synsporos Nann.-Bremek.
On tips of moss leaves on bark of living trees; uncommon.
Recorded from v.c. 59, 60, 65, 69. First record 2008, latest record 2013.
SJ 59; SD 41, 46, 48, 62, 69. 6 sites.

Licea testudinacea Nann.-Bremek.
On bark of living trees; uncommon.
Recorded from v.c. 59, 60, 69. First record 2002, latest record 2014.
SJ 49, 79; SD 39, 46, 61–63, 80. 8 sites.

Licea variabilis Schrad.
On small, decorticated pine branches on the forest floor; common.
Recorded from v.c. 59, 60, 69, 70. First record 1922, latest record 2020.
SD 20, 21, 63; NY 21, 30, 52, 62. 6 sites.

Family Reticulariaceae

Lycogala confusum Nann.-Bremek.
On rotting trunks, especially of beech; uncommon.
59. Ainsdale, 2002, *B. Ing*, SD 21.
69. Dufton Ghyll, 1995, *B. Ing*, NY 62.
70. The Ings, Borrowdale, 1992, *B. Ing*, NY 22; Scales Wood, Buttermere, 1999, *B. Ing*, NY 11.

Lycogala conicum Pers.
On rotting trunks, especially of beech; rare.
59. Duxbury Wood, 2004, *P. Smith*, SD 51.

Lycogala epidendrum agg.
On rotten wood of all kinds, especially in the spring; common. This is an aggregate of two species which were only separated in 1999. In the absence of herbarium specimens older records cannot be assigned to *L. epidendrum sensu stricto* or *L. terrestre*, so all records for the aggregate are listed first.
Recorded from v.c. 59, 60, 65, 69, 70. First record 1859, latest record 2017.
SJ 48, 49, 59, 79, 89; SD 20, 21, 31, 38, 39, 41, 45–52, 54–57, 60–64, 66, 68–71, 73, 81, 90, 91; NY 11, 21, 22, 30, 32, 35, 41, 51, 52, 60–62, 70, 72, 81. 124 sites.

Lycogala epidendrum (L.) Fr.
On rotten wood, especially of conifers; common.
This species has a scarlet plasmodium, dark, rough cortex and a spore mass which varies from grey to olive.
Recorded from v.c. 59, 60, 65, 69, 70. First record 2006, latest record 2019.
SJ 48, 69, 89; SD 20, 41, 46, 52, 71, 73, 68, 90; NY 22, 35, 41. 19 sites.

Systematic Account

Lycogala terrestre Fr.
On rotten wood; common.
This species has pink or cream plasmodium, smooth, pale cortex and pink spore mass.
Recorded from v.c. 59, 60, 65, 69, 70. First record 1961, latest record 2019.
SJ 48, 49, 59, 69; SD 20, 21, 31, 38, 41, 47, 50–52, 60–64, 69, 71, 73, 90; NY 11, 21, 22, 42, 52, 61, 62, 70. 42 sites.

Lycogala flavofuscum Ehrenb.
On dead, standing trees, often in hollows; rare.
69. Kirkby Lonsdale, 1890, **BM**, SD 67.

Reticularia jurana Meylan
Enteridium splendens Rost. var. *juranum* (Meylan) Härkönen
On fallen branches in summer and autumn; common.
Recorded from v.c. 59, 60, 69, 70. First record 1981, latest record 2020.
SD 41, 47, 56, 66; NY 21, 41, 42, 51, 52, 72. 12 sites.

Reticularia liceoides (Lister) Nann.-Bremek.
Enteridium liceoides (Lister) G. Lister
On fallen branches and sticks of conifers; rare.
65. Needle House, 1995, *B. Ing*, SD 79.

Reticularia lobata Lister
Enteridium lobatum (Lister) M.L. Farr
Under the bark of pine stumps, uncommon.
59. Ainsdale, 1972, *B. Ing*, SD 21.

Reticularia lycoperdon Bull.
Enteridium lycoperdon (Bull.) M.L. Farr
On dead standing trees and fallen trunks, especially in the spring; common.
Recorded from v.c. 59, 60, 69, 70. First record 1859, latest record 2019.
SJ 48, 49, 59, 69, 79, 89; SD 20, 21, 31, 38, 39, 41, 45–48, 51, 52, 56, 57, 60–63, 66, 67, 71, 80–83, 90, 91; NY 11, 21, 22, 31, 32, 35, 41, 42, 51, 62, 72, 81. 77 sites.

Reticularia olivacea (Ehrenb.) Fr.
Enteridium olivaceum Ehrenb.
Fallen sticks and branches; uncommon.
69. Skelgill, 1910, *G. Lister*, NY 30.

Tubifera ferruginosa (Batsch) J.F. Gmel.
Tubulifera arachnoidea Jacq.
On rotten conifer trunks in summer; common.
Recorded from v.c. 59, 60, 69, 70. First record 1892, latest record 2018.
SJ 48; SD 20, 21, 31, 38, 39, 41, 47, 51, 61, 66, 71, 90; NY 11, 22, 30, 35, 52, 53, 81. 28 sites.

Systematic Account

Order Trichiales
Family Arcyriaceae

Arcyodes incarnata (Alb. & Schwein.) O.F. Cook
On soggy wood in dried-up ponds and alder carr; rare.
59. Risley Moss, 2018, *J. Watt*, SJ 69.
69. Whinfield Forest, 1986, *B. Ing*, NY 52.
70. Keswick, 1919, *J. Adams*. NY 22.

Arcyria cinerea (Bull.) Pers.
On bark of living trees, especially oak, and on mossy logs; common.
Recorded from v.c. 59, 60, 65, 69, 70. First record 1839, latest record 2012.
SJ 48, 49, 59, 69, 89; SD 20, 21, 38, 39, 47–49, 51, 52, 56, 57, 62, 63, 66, 68, 69; NY 21, 22, 30, 32, 35, 41, 51, 52, 81. 38 sites.

Arcyria denudata (L.) Wettst.
On fallen trunks and stumps of broad-leaved trees; common.
Recorded from v.c. 59, 60, 65, 69, 70. First record 1859, latest record 2019.
SJ 48, 49, 59, 79, 89; SD 20, 21, 38, 39, 45, 47, 51, 56, 61–63, 66, 73, 78, 80, 81; NY 11, 21, 22, 30, 32, 35, 42, 52, 61, 62. 47 sites.

Arcyria ferruginea Sauter
On stumps and fallen trunks, especially in winter; uncommon.
59. Flixton, 1899, *H. Murray*, SJ 79; Formby, 1917, *E.M. Blackwell*, SD 20; Widnes, *B. Hartham*, SJ 58; Ainsdale, 1972, *B. Ing*, SD 21.
70. Carlisle (Massee, 1892) NY 35.

Arcyria incarnata (Pers.) Pers.
On fallen branches, especially of oak; common.
Recorded from v.c. 59, 60, 65, 69, 70. First record 1889, latest record 2018.
SJ 38, 48, 49, 69; SD 20, 21, 38, 39, 41, 47–49, 56, 61–63, 66, 79, 90; NY 11, 21, 22, 30, 31, 35, 41, 42, 51, 52, 62, 63, 70, 72, 81. 41 sites.

Arcyria minuta Buchet
On fallen wood; uncommon.
70. Great Wood, Borrowdale, 2010, *P. Smith*, NY 22.

Arcyria obvelata (Oeder) Onsberg
On fallen trunks and on dead attached branches; common.
Recorded from v.c. 59, 60, 69, 70. First record 1892, latest record 2014.
SJ 48, 79; SD 20, 21, 38, 39, 41, 47, 48, 56, 61, 62, 66; NY 21, 35, 51, 52, 81. 19 sites.

Systematic Account

Arcyria oerstedtii Rostaf.
On stumps and fallen trunks of beech and conifers; uncommon.
59. Widnes, 1957, *B. Hartham*, SJ 58; Stretford, 1963, *B. Hartham*, SJ 89.

Arcyria pomiformis (Leers) Rostaf.
On bark of living trees, especially oak, and on fallen trunks; common.
Recorded from v.c. 59, 60, 65, 69, 70. First record 1889, latest record 2014.
SJ 48, 69, 79, 89; SD 27, 29, 31, 38, 39, 41, 45, 47–49, 51, 56, 61, 63, 69, 71, 73; NY 11, 21, 22, 30–32, 41, 42, 51, 70, 81. 41 sites.

Perichaena chrysosperma (Currey) Lister
On bark of living trees; common.
Recorded from v.c. 59, 60, 65, 69, 70. First record 1983, latest record 2016.
SJ 48, 49, 59, 69, 89; SD 41, 46–49, 51, 57, 61, 62, 66, 67, 69; NY 22, 45. 22 sites.

Perichaena corticalis (Batsch) Rostaf.
Under bark of fallen ash trunks; frequent, but more common in the south.
Recorded from v.c. 59, 60, 69, 70. First record 1892, latest record 2011.
SJ 48; SD 20, 38–40, 47, 56, 66; NY 21, 30, 31, 35, 41, 51, 81. 16 sites.

Perichaena depressa Libert
Under bark of fallen ash trunks; frequent, but more common in the south.
Recorded from v.c. 59, 60, 69, 70. First record 1890, latest record 1992.
SJ 58; SD 20, 60, 66, 67; NY 35. 7 sites.

Perichaena vermicularis (Schwein.) Rostaf.
On deep tree leaf litter; uncommon.
59. Freshfield (Wheldon, 1914) SD 20; Speke, 1976, *B. Ing*, SJ 48.
60. Challon Hall Woods, 1987, *P. Livermore*, SD 47.

Family Dianemataceae

Calomyxa metallica (Berk.) Niewland
On bark of living trees, especially elder; common.
Recorded from v.c. 59, 60, 65, 69, 70. First record 1889, latest record 2013.
SJ 48, 59, 69, 79, 89; SD 39, 41, 46–48, 51, 62, 63, 69, 71; NY 21, 45. 19 sites.

Family Trichiaceae

Hemitrichia abietina (Wigand) G. Lister
On bark of living trees, especially sycamore; rare.
59. Dean Wood, 1994, *NWFG*, SD 67.
70. Applethwaite, 1993, *B. Ing*, NY 22.

Systematic Account

Hemitrichia calyculata (Speg.) M.L. Farr
On fallen trunks, especially of beech; common.
Recorded from v.c. 59, 60, 69, 70. First record 1859, latest record 2019.
SJ 48; SD 20, 38, 47, 51, 66, 67; NY 21, 45. 12 sites.

Hemitrichia clavata (Pers.) Rostaf.
On fallen trunks of broad-leaved trees; uncommon.
60. Long Riddings Wood, Nether Kellet, 1956, *G. Garlick*, SD 56.
70. Carlisle (Massee, 1892) NY 35.

Hemitrichia leiotricha (Lister) G. Lister
On litter under heather; uncommon.
60. Roeburndale, 1984, *B. Ing*, SD 66.

Hemitrichia pardina (Minakata) Ing
Among moss on bark of living trees; uncommon.
59. Bolton, 2006, *P. Smith*, SD 61; Burton Wood, 2008, *B. Ing*, SJ 59.
69. Bowness, 2003, *B. Ing*, SD 49.

Hemitrichia serpula (Scop.) Rostaf.
On compost in hothouse; rare.
70. Carlisle (Massee, 1892) NY 35.
This species, from tropical and warm temperate regions, has recently been found outdoors in Yorkshire, presumably as a response to climate change.

Metatrichia floriformis (Schwein.) Nann.-Bremek.
On fallen wood in secondary woodland; common.
Recorded from v.c. 59, 60, 65, 69, 70. First record 1957, latest record 2019.
SJ 48, 49, 69; SD 20, 21, 38, 39, 41, 45, 47–49, 52, 54–56, 61–63, 66, 68, 71; NY 21, 22, 32, 42, 51, 62. 37 sites.

Metatrichia vesparium (Batsch) Nann.-Bremek.
On fallen trunks, especially of beech and elm; uncommon in the north, more common in the south.
Recorded from v.c. 59, 60, 69, 70. First record 1859, latest record 1984.
SJ 49; SD 45, 47, 48; NY 35. 6 sites.

Oligonema schweinitzii (Berk.) G.W. Martin
On sticks in dried-up ponds and marshes; rare.
59. Ditton Junction, 1963, *J.T. Palmer*, SJ 48.
70. Scotby, Carlisle, 1890, **BM**, NY 35.

Prototrichia metallica (Berk.) Massee
On forest litter, especially of conifers, in winter; rare.
70. Carlisle, 1890, **BM**, NY 35.

Trichia affinis de Bary
On very rotten trunks with moss; common.
Recorded from v.c. 59, 60, 69, 70. First record 1889, latest record 2018.
SJ 69; SD 20, 38, 39, 47–49, 51, 57, 62, 66, 71, 90; NY 11, 21, 22, 30, 35, 41, 42, 51, 81. 30 sites.

Systematic Account

Trichia botrytis (J.F. Gmel.) Pers.
On fallen branches, especially of oak and conifers; common.
Recorded from v.c. 59, 60, 65, 69, 70. First record 1982, latest record 2019.
SJ 48, 69, 79; SD 20, 31, 38, 39, 47–49, 56, 61, 62, 66, 68, 71, 90; NY 11, 21, 22, 30, 31, 35, 41, 42, 51, 52. 53 sites.

Trichia contorta (Ditmar) Rostaf.
On fallen wood; uncommon.
Recorded from v.c. 59, 60, 65, 69, 70. First record 1903, latest record 2010.
SD 31, 38, 39, 47, 48, 69; NY 22, 31. 8 sites.

Trichia decipiens (Pers.) T. Macbr.
On fallen trunks and branches; common.
Recorded from v.c. 59, 60, 65, 69, 70. First record 1889, latest record 2018.
SJ 38, 48, 59, 69, 79; SD 20, 21, 47, 48, 56, 61, 62, 66, 68, 70, 72, 80; NY 11, 21, 22, 32, 35, 42. 35 sites.
Some older records, without herbarium specimens, may be the recently separated *T. meylanii* (see below).

Trichia flavicoma (Lister) Ing
On forest leaf litter; uncommon.
60. Eaves Wood, Silverdale, 1984, *B. Ing*, SD 47.

Trichia lutescens (Lister) Lister
On bark of living trees; uncommon.
59. Levenshulme, 2008, *B. Ing*, SJ 89.
60. Lancaster, 2011, *B. Ing*, SD 46.

Trichia meylanii Ing
On fallen trunks and branches; common.
Recorded from v.c. 59, 60, 69, 70. First record 2011, latest record 2017.
SJ 48, 69; SD 46, 47; NY 45, 52. 9 sites.
This recently separated species differs from *T. decipiens* in its dehiscence via a pre-formed lid and non-reticulated spores; it appears to be as common.

Trichia munda (Lister) Meylan
Among moss on the bark of living trees, especially oak; frequent.
Recorded from v.c. 59, 60, 69, 70. First record 1981, latest record 2012.
SJ 69; SD 18, 46; NY 21, 51, 70. 6 sites.

Trichia persimilis P. Karst.
On fallen trunks, without moss and less rotten; common.
Recorded from v.c. 59, 60, 65, 69, 70. First record 1892, latest record 2012.
SJ 69, 79; SD 38, 39, 47, 48, 56, 57, 61, 66, 68; NY 11, 21, 22, 30, 31, 35, 41, 42, 51, 52, 81. 27 sites.
This is sometimes confused with *T. affinis* but differs in ecology, colour and spore markings.

Systematic Account

Trichia scabra Rostaf.
On rotten wood, especially beech; frequent.
Recorded from v.c. 59, 60, 69, 70. First record 1892, latest record 2019.
SJ 48; SD 21, 47, 51, 52, 60, 61, 71, 90; NY 21, 22, 35. 15 sites.

Trichia varia (Pers.) Pers.
On damp, very rotten wood; common.
Recorded from v.c. 59, 60, 69, 70. First record 1859, latest record 2018.
SJ 48, 49, 58, 68, 69, 79, 89; SD 20, 21, 29, 31, 32, 39–41, 45, 47, 48, 56, 62, 66, 70, 72, 80, 90; NY 11, 21, 22, 31, 35, 41, 42, 51, 52, 62, 70, 72, 81. 44 sites.

Trichia verrucosa Berk.
On fallen branches of conifers; rare.
59. Ainsdale, 2002, *B. Ing*, SD 21.
69. Blea Tarn, 1961, *B. Ing*, NY 20; Roudsea Wood (Frankland, 1966) SD 48.

Order STEMONITIDALES
Family Stemonitidaceae

Amaurochaete atra (Alb. & Schwein.) Rostaf.
On newly felled conifer trunks; uncommon.
59. Freshfield, 1956, *B. Hartham*, SD 20.
60. Challon Hall Woods, 1984, *B. Ing*, SD 47.
69. Whinfell Forest, 1981, *B. Ing*, NY 52.

Brefeldia maxima (Fr.) Rostaf.
On stumps; uncommon.
Recorded from v.c. 59, 60, 69, 70. First record 1859, latest record 2007.
SJ 49; SD 47, 61; NY 52, 53. 7 sites.

Collaria arcyrionema (Rostaf.) Nann.-Bremek.
On mossy stumps in damp woodland; uncommon.
70. Great Wood and The Ings, Borrowdale, 1999, *B. Ing*, NY 22.

Collaria elegans (Racib.) Dhillon & Nann.-Bremek.
On conifer branches on the forest floor; frequent.
59. Mere Sands Wood, 2010, *B. Ing*, SD 41.
60. Alston Hall, 2010, *B. Ing*, SD 63.
69. Lindale, 2011, *B. Ing*, SD 48; Lowther Park, 2011, *B. Ing*, NY 52.

Collaria rubens (Lister) Nann.-Bremek.
On holly leaf litter; rare.
59. Anglezarke, 1972, *B. Ing*, SD 61.

Colloderma oculatum (Lippert) G. Lister
On mosses and lichens on the bark of living trees; uncommon.
Recorded from v.c. 59, 60, 69, 70. First record 1976, latest record 1995.
SJ 48; SD 38, 47, 48, 66; NY 11, 21, 41, 51, 62, 81. 12 sites.

Systematic Account

Comatricha alta Preuss
On old logs, especially on the cut end; uncommon.
59. Flixton, 1899, *H. Murray*, SJ 79.

Comatricha laxa Rostaf.
On dead wood and, occasionally, on the bark of living trees; frequent.
Recorded from v.c. 59, 60, 69, 70. First record 1919, latest record 2010.
SD 39, 41, 61, 63; NY 21, 22. 8 sites.

Comatricha nigra (Pers.) Schröt.
On dead wood of all kinds; common.
Recorded from v.c. 59, 60, 65, 69, 70. First record 1859, latest record 2018.
SJ 48, 49, 59, 69, 79, 89; SD 20, 21, 29, 38, 39, 41, 47–49, 51, 56, 61, 66, 68, 71, 90; NY 11, 21, 22, 30, 31, 32, 35, 41, 42, 45, 51, 62, 72, 81. 54 sites.

Comatricha pulchella (C. Bab.) Rostaf.
On leaf litter, especially of holly; common.
Recorded from v.c. 59, 60, 69, 70. First record 1922, latest record 2003.
SJ 48; SD 21, 29, 38, 47, 48, 56, 66; NY 11, 21, 22, 41, 42, 51. 14 sites.

Comatricha tenerrima (M.A. Curtis) G. Lister
On dead herbaceous stems in damp sites; uncommon.
59. Dean Wood, 1994, *NWFG*, SD 61.
60. Lancaster, 2011, *B. Ing*, SD 46.
69. Flakebridge, 1995, *B. Ing*, NY 62.
70. The Ings, Borrowdale, 1992, *B. Ing*, NY 22; Thornthwaite Forest, 1999, *B. Ing*, NY 22.

Diacheopsis insessa (G. Lister) Ing
On lichens on the bark of living trees; uncommon.
69. Hartsop, 1981, *B. Ing*, NY 41; Nibthwaite, 1981, *B. Ing*, SD 29.
70. Borrowdale, 1981, *B. Ing*, NY 21; Scales Wood, Buttermere, 1992, *B. Ing*, NY 11.

Enerthenema papillatum (Pers.) Rostaf.
On bark of living trees and on fallen branches; common.
Recorded from v.c. 59, 60, 69, 70. First record 1889, latest record 2013.
SJ 48, 49, 69, 79; SD 21, 29, 38, 39, 41, 47–51, 56, 57, 63, 66, 72; NY 11, 21, 22, 35, 42, 51, 52, 72, 81. 25 sites.

Lamproderma columbinum (Pers.) Rostaf.
On mosses on logs and rocks in damp woodland; frequent.
Recorded from v.c. 59, 60, 69, 70. First record 1892, latest record 2018.
SJ 89; SD 31, 48, 66; NY 11, 21, 32, 35, 41, 42, 51, 52. 13 sites.

Systematic Account

Lamproderma nigrescens (Rostaf.) Rostaf.
Lamproderma arcyrioides auctt.
On forest leaf litter; uncommon.
59. Crosby, 1909, *J.W. Ellis*, SD 30.
60. Gait Barrows, 1987, *P. Livermore*, SD 47.
69. Tilberthwaite, 1963, *B. Ing*, NY 30.
70. Carlisle (Massee, 1892) NY 35.

Lamproderma scintillans (Berk. & Broome) Morgan
On fern and leaf litter, especially of holly; common.
Recorded from v.c. 59, 60, 69, 70. First record 1909, latest record 2011.
SJ 48, 79; SD 31, 38, 47, 48, 56, 66; NY 45, 72. 12 sites.

Macbrideola cornea (G. Lister & Cran) Alexop.
On mosses on the bark of living trees; common.
Recorded from v.c. 59, 60, 65, 69, 70. First record 1981, latest record 2013.
SD 29, 41, 48, 59, 66, 69, 73; NY 21, 22, 31, 41, 42, 51. 15 sites.

Macbrideola macrospora (Nann.-Bremek.) Ing
On bark of living trees, not usually on moss; uncommon.
65. Sedbergh, 2010, *B. Ing*, SD 69.
69. Lindale, 2011, *B. Ing*, SD 48.

Paradiacheopsis cribrata Nann.-Bremek.
On bark of living trees; frequent.
Recorded from v.c. 59, 60, 65, 69, 70. First record 1981, latest record 2013.
SD 29, 38, 49, 50, 60, 63, 69, 73; NY 21, 22, 31, 42, 44. 16 sites.

Paradiacheopsis fimbriata (G. Lister & Cran) Hertel
On acid bark of living trees; common.
Recorded from v.c. 59, 60, 69, 70. First record 1980, latest record 2011.
SJ 33, 38, 47–49, 56, 59, 69; SD 21, 38, 39, 41, 48–50, 71; NY 22, 42, 44, 52, 62, 72. 27 sites.
This species appears to be tolerant of urban air pollution and is usually associated with the green alga *Desmococcus olivaceus*.

Paradiacheopsis longipila Hoof & Nann.-Bremek.
On bark of living oak; rare.
70. Southwaite, 2008, *B. Ing*, NY 44.
This is the first British record, it has subsequently been found in Banff, Scotland. It seems to be a northern species in Europe.

Paradiacheopsis microcarpa (Meylan) D.W. Mitchell
On acid bark of living trees; uncommon.
59. Cuerden Valley, 2009, *P. Smith*, SD 52.

Systematic Account

Paradiacheopsis rigida (Brandza) Nann.-Bremek.
On bark of living trees; uncommon.
59. Borsdane Wood, 2009, *P. Smith*, SD 60; Rivington Country Park, 2010, *P. Smith*, SD 61.

Paradiacheopsis solitaria (Nann.-Bremek.) Nann.-Bremek.
On bark of living trees; common.
Recorded from v.c. 59, 60, 65, 69, 70. First record 1981, latest record 2013.
SJ 49, 69; SD 29, 38, 39, 47, 49–51, 56, 57, 66, 69; NY 11, 21, 22, 41, 42, 51, 70, 72. 30 sites.

Stemonitis axifera (Bull.) T. Macbr.
On fallen trunks; common.
Recorded from v.c. 59, 60, 69, 70. First record 1892, latest record 2018.
SJ 89; SD 22, 38, 39, 45, 47, 48, 51, 56, 66, 71, 80; NY 11, 21, 35, 41, 42, 51, 52, 62. 23 sites.

Stemonitis flavogenita E. Jahn
On fallen trunks; common.
Recorded from v.c. 59, 60, 69, 70. First record 1918, latest record 1992.
SJ 48; SD 20, 21, 38, 40, 45, 66; NY 21, 22, 52, 62. 12 sites.

Stemonitis fusca Roth
On fallen trunks; common.
Recorded from v.c. 59, 60, 65, 69, 70. First record 1889, latest record 2019.
SJ 48; SD 20, 21, 30, 38, 39, 41, 47–49, 51, 56, 61, 62, 66, 70, 79, 90; NY 11, 21, 22, 30–32, 35, 41, 42, 51, 52, 62, 70, 81. 46 sites.

Stemonitis herbatica Peck
On forest leaf litter; uncommon.
59. Cuerden Valley, 2011, *P. Smith*, SD 52.
70. The Ings, Borrowdale, 1999, *B. Ing*, NY 22.

Stemonitis nigrescens Rex
On bark of living trees; uncommon.
59. Healy Dell, 2000, *NWFG*, SD 81.
60. Eaves Wood, Silverdale, 1981, *B. Ing*, SD 47.
This species is often united with *S. fusca* but is confined to tree bark rather than fallen wood. It is smaller, darker and with slightly larger spores. The ecological difference is enough to maintain its separation.

Stemonitis smithii T. Macbr.
On small branches and twigs in very damp woodland; rare.
70. Scales Wood, Buttermere, 1992, *B. Ing*, NY 11.

Systematic Account

Stemonitis splendens Rostaf.
On fallen trunks, usually in Atlantic woodland; rare.
59. Scutcher's Acres, 2018, *J. Watt*, 2018.
60. Potts Wood, Warton, 1990, *P. Livermore*, SD 47.
This species occurs in Britain mainly along the west coast and the more Atlantic parts of the south coast; it is common in the tropics.

Stemonitopsis amoena (Nann.-Bremek.) Nann.-Bremek.
On the bark of living trees; uncommon.
59. Ainsdale, 2002, *B. Ing*, SD 21; Rufford Old Hall, 2006, *B. Ing*, SD 41.

Stemonitopsis hyperopta (Meylan) Nann.-Bremek.
On very rotten conifer trunks; frequent.
59. Ainsdale, 1988, *B. Ing*, SD 31 and 2002, *B. Ing*, SD 21.
60. Alston Hall, 2010, *B. Ing*, SD 63.
69. Whinfell Forest, 1981, *B. Ing*, NY 52; Lowther Park, 2011, *B. Ing*, NY 52.

Stemonitopsis reticulata (H.C. Gilbert) Nann.-Bremek. & Y. Yamam.
On rotten conifer wood; rare.
59. Anglezarke, 1972, *B. Ing*, SD 61. This was the first British record; it has since been found in Denbighshire and Derbyshire.

Stemonitopsis typhina (F.H. Wigg) Nann.-Bremek.
On wet, very rotten wood; common.
Recorded from v.c. 59, 60, 69, 70. First record 1892, latest record 2011.
SD 20, 21, 38, 39, 47, 51, 56, 61, 66; NY 21, 22, 31, 35, 51, 52, 62. 21 sites.

Symphytocarpus amaurochaetoides Nann.-Bremek.
On fallen trunks and stumps; uncommon.
59. Ainsdale, 2002, *B. Ing*, SD 21.

Symphytocarpus flaccidus (Lister) Ing & Nann.-Bremek.
On dead standing pine trunks; common.
59. Ainsdale, 1972, 2002, *B. Ing*, SD 21.
60. Eaves Wood, Silverdale, 1979, *P. Livermore*, SD 47.
69. Lowther Park, 1981, 2011, *B. Ing*, NY 52.

Symphytocarpus impexus Ing & Nann.-Bremek.
In twiggy litter on the forest floor; uncommon.
59. Bowkers Green, 1917, **BM**, SD 40.

Systematic Account

Order PHYSARALES
Family Didymiaceae

Diachea leucopodia (Bull.) Rostaf.
On leaf litter and living herbaceous stems; common in the south, rare in the north.
59. Freshfield, 1960, *S.S. Bates*; 1965, *B. Ing*, SD 21.
60. Hawes Water, 2003, *B. Ing*, SD 47.
69. Dufton Ghyll, 1995, *B. Ing*, NY 52.
70. Carlisle (Massee, 1892) NH 35.

Diachea subsessilis Peck
On vegetable litter in damp places; rare.
59. Roddlesworth, 1991, *P. Smith*, SD 62.

Diderma chondrioderma (de Bary & Rostaf.) G. Lister
On moss on the bark of living trees; frequent.
Recorded from v.c. 59, 60, 69, 70. First record 1997, latest record 2014.
SD 41, 46, 63, 80; NY 22, 60. 6 sites.

Diderma cinereum Morgan
On terrestrial woodland mosses; rare.
59. Duxbury Wood, 2004, *NWFG*, SD 51; Risley Moss, 2012, *P. Smith*, SJ 69.
70. Great Wood, Borrowdale, 2010, *P. Smith*, NY 22.

Diderma deplanatum Fr.
On leaf litter and terrestrial woodland mosses; uncommon.
60. Gait Barrows, 1985, *P. Livermore*, SD 47.
69. Roudsea Wood, 2002, *B. Ing*, SD 38.
70. Carlisle (Massee, 1892) NY 35; Thirlmere, *BMSF*, NY 31.

Diderma donkii Nann.-Bremek.
On forest leaf litter; uncommon.
69. Roudsea Wood, 2002, *B. Ing*, SD 38.

Diderma effusum (Schwein.) Morgan
On leaf litter; frequent.
59. Whalley, 1952, *S.S. Bates*, SD 73; Liverpool, 2018, *A. Carter*, SJ 48.
60. Gait Barrows, 1985, *P. Livermore*, SD 47.
65. Sedbergh, 1971, *P. Holland*, SD 68.
70. Latrigg, 1996, *B. Ing*, NY 22.

Diderma floriforme (Bull.) Pers.
On fallen branches in ancient woodland, especially on oak; uncommon.
70. Thornthwaite Forest, 1999, *B. Ing*, NY 22.

Diderma globosum Pers.
On vegetation in marshy places; uncommon.
59. Ainsdale, 2002, *B. Ing*, SD 21.
69. Naddle Low Forest, 1981, *B. Ing*, NY 51; Roudsea Wood, 2002, *B. Ing*, SD 38.

Systematic Account

Diderma hemisphaericum (Bull.) Hornem.
On leaf litter in damp places; frequent.
Recorded from v.c. 59, 60, 69, 70. First record 2005, latest record 2019.
SJ 48; SD 61, 63, 68; NY 45. 6 sites.

Diderma lucidum Berk. & Broome
On moss on boulders and in ravines in areas of high rainfall; rare.
70. Lodore Falls, Borrowdale, 1981, *B. Ing*, NY 21; Scales Wood, Buttermere, 1981, *B. Ing*, NY 11.
This the typical, and very beautiful, key member of the ravine myxomycete association (Ing, 1983).

Diderma meyerae H. Singer, G. Moreno, C. Illana & A. Sanchez
Diderma niveum auctt.
On vegetation at the edge of melting snow on mountains in spring; uncommon.
69. Helvellyn, 1997, *B. Ing*, NY 31.
70. Skiddaw, 1997, *B. Ing*, NY 22; Cross Fell, 1997, *B. Ing*, NY 63.

Diderma ochraceum Hoffm.
On mosses on boulders and in ravines; uncommon.
70. Aira Force, 1983, *B. Ing*, NY 32; Scales Wood, Buttermere, 1981, 1999, *B. Ing*, NY 11; Johnny Wood, Borrowdale, 1981, *B. Ing*, NY 21.

Diderma spumarioides (Fr.) Fr.
In dune and limestone grassland; uncommon.
59. Freshfield, 1987, *B. Ing*, SD 20; Ainsdale, 2002, *B. Ing*, NY 21; Hightown, 2003, *B. Ing*, SD 20.
60. Gait Barrows, 1985, *P. Livermore*; 2002, *B. Ing*, SD 47.
70. Ravenglass, 2011, *B. Ing*, SD 08.

Diderma umbilicatum Pers.
On twiggy litter, especially of brambles; uncommon.
59. Manchester (Travis, 1925) SJ 89; Ainsdale, 2010, *P. Smith*, SD 21.
60. Gait Barrows, 1987, *P. Livermore*, SD 47.
70. Scales Wood, Buttermere, 1981, *B. Ing*, NY 11.

Didymium anellus Morgan
On leaf litter; uncommon.
69. Flakebridge, 1995, *B. Ing*, NY 62.

Didymium bahiense Gottsb.
On herbaceous litter; common.
Recorded from v.c. 59, 60, 65, 69, 70. First record 1965, latest record 2018.
SJ 89; SD 39, 45, 69; NY 45, 62. 6 sites.

Systematic Account

Didymium clavus (Alb. & Schwein.) Rabenh.
On leaf litter; common.
59. Speke Hall, Liverpool, 1972, 1999, *B. Ing*, SJ 48; Levenshulme, 2018, *B. Ing*, SJ 89.
60. Hawes Water, 2003, *B. Ing*, SD 47.
69. Murton, 1981, *B. Ing*, NY 72.
70. Carleton, 2011, *B. Ing*, NY 45.

Didymium crustaceum Fr.
On old Indian cloth, usually on litter in hedge bottoms; rare.
59. Trafford Park, 1960, *J.T. Palmer*, SJ 79.

Didymium difforme (Pers.) S.F. Gray
On plant litter of all kinds; common.
Recorded from v.c. 59, 60, 65, 69, 70. First record 1889, latest record 2015.
SJ 38, 39, 48, 49, 69, 79, 89; SD 20, 30, 38, 39, 47, 49, 50, 56, 57, 66, 68; NY 21, 22, 30, 31, 35, 41, 42, 51, 52, 70, 72, 81. 47 sites.

Didymium ilicinum Ing
On leaf litter, especially of holly; common.
59. Risley Moss, 2012, *B. Ing*, SJ 48.
60. Lancaster, 2011, *B. Ing*, SD 46.
69. Eggerslack Woods, 2011, *B. Ing*, SD 47.
70. Carleton, 2011, *B. Ing*, NY 45.
This is a recent segregate from *D. squamulosum* (see below), which is an aggregate of non-sexual species.

Didymium megalosporum Berk. & M.A. Curt.
On old Indian cloth, its usual habitat is plant waste; uncommon.
59. Trafford Park, 1960, *J.T. Palmer*, SJ 79; Winston, 2018, *A. Carter*, SJ 48; Liverpool, 2018, *A. Carter*, SJ 48.

Didymium melanospermum (Pers.) T. Macbr.
On conifer litter and wood, and also moss stems on acid soil; common.
Recorded from v.c. 59, 60, 69, 70. First record 1836, latest record 1981.
SJ 48, 49, 89; SD 39, 57; NY 21, 30, 31, 35, 52. 10 sites.

Didymium minus (Lister) Morgan
On dead herbaceous stems; frequent.
59. Speke Hall, 1972, *B. Ing*, SJ 48; Bolton, 2011, *P. Smith*, SD 61.
60. Gait Barrows, 2002, *B. Ing*, SD 47.
69. Bowness (Atkinson, 1889) SD 49.

Didymium nigripes (Link) Fr.
On leaf litter, especially of holly; common.
Recorded from v.c. 59, 60, 69, 70. First record 1851, latest record 2010.
SJ 58, 59, 79; SD 38, 39, 41, 45, 47, 51, 56; NY 21, 30, 35. 15 sites.

Systematic Account

Didymium serpula Fr.
On forest leaf litter; uncommon.
59. Freshfield, 1960, *S.S. Bates*, SD 20.

Didymium squamulosum (Alb. & Schwein.) Fr.
On leaf litter of all kinds; common.
Recorded from v.c. 59, 60, 65, 69, 70. First record 1898, latest record 2019.
SJ 38, 39, 48, 49, 59, 68, 69, 79, 89; SD 20, 21, 38, 39, 45, 47, 48, 56, 57, 61, 66, 70, 78; NY 21, 22, 30, 32, 35, 41, 42, 45, 51, 52, 72, 81. 48 sites.
This is now known to be a complex of non-sexual species, the next most common being *D. ilicinum* (see above).

Didymium verrucosporum A.L. Welden
On herbaceous litter, uncommon.
59. Flixton 1906, *H. Murray*, SJ 79.

Lepidoderma tigrinum (Schrad.) Rostaf.
On mosses on wet boulders, in ravines and on moss-covered fallen conifer trunks; frequent in the high rainfall areas.
Recorded from v.c. 60, 69, 70. First record 1922, latest record 1983.
SD 66; NY 11, 21, 32, 42, 51, 52. 8 sites.

Lepidoderma trevelyanii (Grev.) Poulain & Mar. Mey.
Diderma trevelyanii (Grev.) Fr.
On alder litter; rare.
59. Freshfield, 1918, *E.M. Blackwell*, SD 20.

Mucilago crustacea F.H. Wigg.
On living grasses, especially on basic soils; common.
Recorded from v.c. 59, 60, 69, 70. First record 1859, latest record 2018.
SJ 39, 48, 49, 89; SD 18, 20, 21, 31, 32, 39, 45, 47, 48, 52, 56, 61, 71, 91; NY 35, 52, 62. 21 sites.
This is the true 'dog vomit slime mould' a name wrongly given to *Fuligo septica* (see below).

Family Physaraceae

Badhamia affinis Rostaf.
On mosses on bark of living trees; frequent, especially in the west.
Recorded from v.c. 59, 60, 65, 69, 70. First record 1981, latest record 2014.
SJ 48, 89; SD 41, 46, 47, 49, 61, 69; NY 22, 62. 10 sites.

Badhamia foliicola Lister
On living grasses; uncommon.
59. Bolton, 2006, *P. Smith*, SD 61.

Systematic Account

Badhamia gracilis (T. Macbr.) T. Macbr.
On tobacco waste; extinct in Britain.
59. Formby, 1959, *S.S. Bates*, SD 20.
This is essentially an American species which grows on decaying cacti, especially *Opuntia*. It is now found in the Mediterranean region, North Africa and the Canary Islands, on cacti introduced to support the cochineal bug. Its casual appearance in Lancashire was no doubt associated with the import and processing of tobacco. The species has been confused with *B. melanospora*, a rare species on wood, which occurs in a few sites in southern England.

Badhamia lilacina (Fr.) Rostaf.
On vegetation growing on *Sphagnum* bogs; uncommon.
69. Meathop Moss, 1961, *B. Ing*, SD 48; Roudsea Moss, 2002, *B. Ing*, SD 38.
70. Carlisle (Massee, 1892) NY 35.
This inconspicuous species is characteristic of the mosslands of the North West and Cheshire but is very difficult to spot, unless in its egg yellow, aquatic plasmodial stage. It is likely to be found more often with careful searching.

Badhamia macrocarpa (Ces.) Rostaf.
On moss on bark of living trees, on branches rather than trunks, and occasionally on fallen wood; uncommon.
Recorded from v.c. 59. First record 2004, latest record 2019.
SJ 48; SD 31, 61, 62. 6 sites.

Badhamia nitens Berk.
On bark of living trees; rare.
70. Carlisle (Massee, 1892) NY 35.

Badhamia panicea (Fr.) Rostaf.
On fallen trunks, especially of beech and elm; common.
Recorded from v.c. 59, 60, 69, 70. First record 1892, latest record 1981.
SJ 49, 69, 79; SD 47, 52; NY 35, 52, 72. 9 sites.

Badhamia utricularis (Bull.) Berk.
On fungi such as *Stereum hirsutum* and *Phlebia radiata* on fallen trunks, especially in winter; common.
Recorded from v.c. 59, 60, 69, 70. First record 1900, latest record 2010.
SJ 39, 48, 69, 89; SD 21, 38, 47, 49; NY 35, 51, 52, 62. 12 sites.

Systematic Account

Craterium aureum (Schumach.) Rostaf.
On leaf litter, especially of beech; uncommon.
59. Freshfield, 1960, 1965, *S.S. Bates*, SD 20.
70. Carlisle (Massee, 1892) NY 35.

Craterium leucocephalum (Pers.) Ditmar
On leaf litter; uncommon.
59. Formby, 1909, *H. Wheldon*, SD 20.
50. Hawes Water, 2003, *B. Ing*, SD 47.
69. Stenkrith Park, 1981, *D.W. Mitchell*, NY 70; Dufton Ghyll, 1986, *B. Ing*, NY 62.
70. Carlisle (Massee, 1892) NY 35.

Craterium minutum (Leers) Fr.
On leaf litter and living herbaceous stems; common.
Recorded from v.c. 59, 60, 69, 70. First record 1889, latest record 2012.
SJ 48, 59, 79, 89; SD 20, 21, 38, 39, 42, 47–49, 56, 61, 66, 81, 80; NY 21, 22, 30, 35, 41, 51, 62, 81. 32 sites.

Craterium muscorum Ing
On mosses on wet boulders and in ravines; uncommon.
Recorded from v.c. 69, 70. First record 1892, latest record 1981.
SD 29, 39, 48; NY 21, 30, 32, 35. 9 sites.

Fuligo candida Pers.
On fallen wood; uncommon.
59. Liverpool, 2017, *A. Carter*, SJ 48.
69. Brantwood, Coniston, 1961, *B. Ing*, SD 39; Tarn Hows, 1963, *B. Ing*, SD 39.

Fuligo muscorum Alb. & Schwein.
On terrestrial woodland mosses, on wet mossy rocks and in ravines; frequent in the west.
Recorded from v.c. 69, 70. First record 1892, latest record 2002.
SD 38, 39; NY 11, 21, 22, 35. 6 sites.

Fuligo rufa Pers.
On very rotten wood and sawdust heaps, often in exposed, warm sites; uncommon.
59. Formby, 1959, *S.S. Bates*, SD 20.
60. Gait Barrows, 2002, *B. Ing*, SD 47.
69. Bowness, 2003, *B. Ing*, SD 49.

Systematic Account

Fuligo septica (L.) F.H. Wigg.
On fallen wood of all kinds and, historically, on tan heaps; common.
Recorded from v.c. 59, 60, 65, 69, 70. First record 1859, latest record 2018.
SJ 48, 58, 59; SD 20, 21, 25, 31, 38, 39, 41, 45, 47, 48, 51, 52, 55–57, 60–63, 66, 70, 71, 73, 79, 90, 91; NY 11, 21, 22, 31, 32, 41, 42, 51, 52, 55, 62, 70, 72, 81. 55 sites.
This species is the traditional 'Flowers of Tan' but in several recent books it has been labelled as 'Dog vomit slime mould' which, of course, is a grassland species, *Mucilago crustacea*.

Leocarpus fragilis (Dicks.) Rostaf.
On conifer litter and climbing stems of herbaceous and woody plants; common.
Recorded from v.c. 59, 60, 69, 70. First record 1859, latest record 2011.
SJ 48, 88, 89; SD 20, 21, 31, 38, 39, 45, 47, 48, 50, 56, 66, 90; NY 11, 21, 22, 30, 31, 35, 41, 51. 30 sites.

Physarum album (Bull.) Chevall
Physarum nutans Pers.
On fallen wood of all kinds; common.
Recorded from v.c. 59, 60, 69, 70. First record 1889, latest record 2017.
SJ 48, 49, 69, 79; SD 20, 21, 38, 39, 41, 47–49, 51, 52, 56, 57, 61, 66, 90; NY 11, 21, 22, 30, 31, 35, 41, 42, 51, 52, 61, 70, 72, 81. 38 sites.

Physarum auriscalpium Cooke
On mossy bark of living trees; frequent.
59. Risley Moss, 2008, *B. Ing*, SJ 69.
60. Outhwaite Wood, 1976, *D. Galvin*, SD 66.

Physarum bethelii T. Macbr.
On mossy bark of living trees; uncommon.
60. Lord's Lot Wood, 1984, *B. Ing*, SD 57.

Physarum bitectum G. Lister
On litter under brambles; uncommon.
60. Gait Barrows, 1987, *P. Livermore*, SD 47; Heysham Nature Reserve, 1990, *P. Livermore*, SD 45.

Physarum bivalve Pers.
On leaf litter; common.
Recorded from v.c. 59, 60, 69, 70. First record 1892, latest record 2010.
SJ 48, 49, 59, 99; SD 20, 21, 45, 47, 57; NY 35. 10 sites.

Physarum cinereum (Batsch) Pers,
On grass litter and lawns; common.
Recorded from v.c. 59, 60, 69, 70. First record 1892, latest record 2019.
SD 20, 41, 48, 61, 63; NY 21, 32, 35. 89 sites.

Systematic Account

Physarum citrinum Schumach.
On mossy wood and terrestrial woodland mosses; rare.
70. Carlisle (Massee, 1892) NY 35; Keswick, 1910, *G. Lister*, NY 22; Johnny's Wood, Borrowdale, 1992, *B. Ing*, NY 21.

Physarum compressum Alb. & Schwein.
On herbaceous litter; common.
Recorded from v.c. 59, 60. First record 1909, latest record 2003.
SJ 48, 58, 69, 79; SD 20, 47, 61, 80. 11 sites.

Physarum conglomeratum (Fr.) Rostaf.
On leaf and moss litter; rare.
60. Gait Barrows, 1985, *P. Livermore*, SD 47.

Physarum contextum (Pers.) Pers.
On leaf and moss litter; rare.
59. Freshfield, 1965, *B. Ing*, SD 20.
60. Lytham Hall, 2018, *J. Watt*, SD 32.

Physarum crateriforme Petch
On bark of living trees; uncommon.
69. Witherslack Woods, 1981, *B. Ing*, SD 48; Bowness, 2003, *B. Ing*, SD 49.
70. Scales Wood, Buttermere, 1981, *B. Ing*, NY 11.

Physarum decipiens M.A. Curtis
On mossy bark of living trees; uncommon.
59. Carr Mill Dam, 2011, *P. Smith*, SJ 59.

Physarum didermoides (Pers.) Rostaf.
On straw heaps; once common, now rare.
70. Carlisle (Massee, 1892) NY 35.

Physarum globuliferum (Bull.) Pers.
On rotten wood; once frequent, now rare.
69. Blea Tarn, 1961, *B. Ing*, NY 20.
70. Keswick, 1919, *A. Adams*, NY 21.

Physarum leucophaeum Fr.
On fallen branches; common.
Recorded from v.c. 59, 60, 69, 70. First record 1898, latest record 2018.
SJ 48, 59, 79; SD 20, 21, 31, 38, 39, 41, 45, 47, 48, 51, 56, 57, 61, 62, 66; NY 11, 21, 35, 42, 51, 52, 81. 28 sites.

Physarum leucopus Link.
On leaf litter; uncommon.
70. Carlisle (Massee, 1892) NY 35.
This a lookalike to *Didymium squamulosum*, until it is seen under the microscope, so may have been overlooked.

Systematic Account

Physarum limonium Nann.-Bremek.
On bark of living trees; rare.
70. Gowbarrow Park, 1981, *B. Ing*, NY 42.

Physarum murinum Lister
On rotten conifer wood; now extinct in Britain.
70. Keswick, 1919, *A. Adams*, NY 22.

Physarum psittacinum Ditmar
On rotten trunks in ancient woodland; uncommon.
70. Carlisle (Massee, 1892) NY 35.

Physarum pusillum (Berk. & M.A. Curt.) G. Lister
On grass litter of all kinds and occasionally on the bark of living trees; common.
Recorded from v.c. 59, 60, 69. First record 1906, latest record 2018.
SJ 48, 79; SD 17, 20, 21, 40, 46, 60, 61. 89 sites.

Physarum robustum (Lister) Nann.-Bremek.
On fallen trunks and branches; common.
Recorded from v.c. 59, 60, 69, 70. First record 1987, latest record 2010.
SJ 39, 48, 59; SD 21, 38, 47, 63; NY 42. 9 sites.

Physarum straminipes Lister
On straw heaps and grass litter; now rare.
59. Thornton, 1925, *A. Richardson*, SD 30.

Physarum virescens Ditmar
On terrestrial woodland feather mosses; frequent in the west.
Recorded from v.c. 59, 69, 70. First record 1892, latest record 2018.
SD 38, 39, 66; NY 11, 21, 35, 42, 51, 52, 60. 11 sites.

Physarum viride (Bull.) Pers.
On conifer brashings and small, fallen branches, especially of oak; frequent.
Recorded from v.c. 59, 60, 69, 70. First record 1859, latest record 1981.
SJ 48, 49; SD 20, 38–40, 48, 57, 61; NY 21, 22, 35, 52. 14 sites.

ILLUSTRATIONS

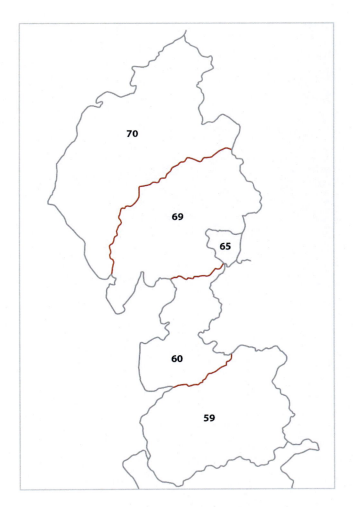

Figure 1: The vice-counties of Lancashire and Cumbria

The Slime Moulds of Lancashire and Cumbria

Figure 2: *Arcyodes incarnata* – an uncommon species on sticks in damp places (John Watt)

Illustrations

Figure 3: *Arcyria obvelata* – a common species on fallen trunks, especially beech (John Robinson)

Figure 4: *Lepidoderma tigrinum* – an uncommon species of wet, mossy rocks and moss and lichen on rotting conifer trunks (Michel Poulain)

Illustrations

Figure 5: *Licea biforis* – a relative newcomer to our islands, arriving from the tropics as a harbinger of climate change (John Robinson)

Figure 6: *Macbrideola cornea* – a common species, especially in the west, on mosses on the bark of living trees (Diana Wrigley-Basanta)